AN ENGINEERING

VIEW

of the UNIVERSE

VOL IV –

GRAVITY and MORE

Robert Heilman

The Series

AN ENGINEERING VIEW OF THE

UNIVERSE

Books from the point of view of Engineering Methodology.

VOL I – So many things wrong, the groundwork.

VOL II – A solution for PI, just because Engineers CAN. More questions about the Universe.

VOL III – ENOUGH questions. Tying all things together and why. A Unified version of the Universe.

VOL IV – Completing the Unified Theory in more detail, especially Gravity. Last Book, UNLESS....

Dedication Page

After Unifying the Universe there are many people that I need to thank, because let's face it, I never could have done this alone. I'm not that smart!

TO,

RJ, my research assistant, my sanity check, my Son, my U of M student. Wouldn't be as good without your Critique.

Danielle, Thanks for all the Stars. My free thinker!

Amanda, Thanks for the Brain Trust, a place to think and work, the battery for my computer and the Maricopa County Library.

Rachael, Thanks for the Ypsilanti Beer Festival. Nothings helps understand Physics like a Beer Festival! Or a 100 other places we've done.

Jake, it all starts and ends with you. You are my Universe. Someday you may even stop looking at me funny!

Zoe and Millie, My Brain Trust. Your understanding of what makes things move (treats) in the Universe is Amazing.

COPYRIGHT PAGE

<u>TABLE OF CONTENTS</u>

INTRODUCTION

One of the questions that I get is, Why write a book on the Universe? I am an Engineer, not a Physicist. Nor do I have a great knowledge of the Universe or considered writing a book about the Universe. But what I do have, as most all Engineers, is the methodology to figure out how things work and why. Well one day I decided to read about Einstein and SpaceTime. Sounded pretty complicated but what the heck. I'm reading and Einstein writes that Gravity can distort SpaceTime. I don't know much, but I have read that Space is thought to be mostly empty. So how does Gravity distort empty Space? I search: Videos, Books, Articles, et al. But not one explanation on how this is possible. This is the basis for this whole Theory, with no explanation! Without the details, this is just a good Story. So this amazes me that there is no reference to the mechanics, or how things move. I read more Stories; Big Bang, BlackHoles, Gravitational waves, Speed of Light, etc. Still no mechanics. And since nobody seems to care why things happen, I decided to give my Engineering opinion of why some things could or could not happen. Now is it just an educated guess? Yes, but so is every other Theory that has no Physical Proof or the mechanics of how it could happen, Einstein included. I topic led to another and another and another. Then I began to see commonalities, and one night I said this can all be Unified. So there you have it, the Universe according to an Engineer. So every Physicist that has a Unified Theory of the Universe, please stand over hear on this side of the room with me. It sure is getting lonely here. And see what understanding mechanics can do?

THE BEGINNINGS

Sorry, the Title for this book, VOL IV, was supposed to be
IRRELAVENT IMPORTANT STUFF, but that would not fit on
the Cover. So I had to name the book GRAVITY. Worse yet, I
then had to explain Gravity, the only word that fits on the Cover!
I should have named it "PIGS CAN FLY" but oh well Ok, the
book. Never thought I would write a book, especially about the
state of Physics and the Universe, But, you never know. My
name is Robert Heilman. When I graduated from High School I
actually thought I was going to be an Accountant. So off to
College I went and began learning Accounting. I was introduced
to computers that were being integrated into the Business world.
I learned Programming and computer code was like a breath of
fresh air compared to Accounting. I loved the creativity and
clear output results. To support myself while attending school, I
landed a job at a local automotive parts manufacturer. Because
Computers were starting to lead robotics Into manufacturing, I
found myself being pulled into Programming more and more.
Getting my hands dirty, creativity, working with numbers,
precision, and measureable results, I thought this was IT. But,
as things go, the company was bought out and the staff reduced.
I found myself without a job. So I began looking, But I was
convinced that Accounting was out. The type of jobs I was

looking for were far and few between, especially in my small Town. 60 miles away, in a city called Detroit, jobs were plentiful. Because of my computer and programming experience. I landed a job as an NC programmer and Design Studio support for an Automotive supplier. I was exposed to many things I never knew existed! Computer Aided Design, Computer Aided Simulations, Crash worthiness, weight reduction, Aerodynamics and wind tunnel testing, Strength of materials, Road Testing, In Process Testing, Analyzing Reports, Writing Reports, and many others; quality, design, production, inspection, cost reduction methodologies from the US, Japan, Germany and around the world. Things such as Statistical Process Control, Six Sigma, Etc. This was Engineering! It was a natural fit, especially with the integration Of computers. Within a few years I was an Engineer And never looked back. And now Space is calling. But differently than it calls Physicists. I don't care about Galaxies or other things 100 million light years away and I don't care about things that can't be proved. So, no, I don't care about Higgs bosons or Black Holes. I am an Engineer. Give me the test data. Prove it exists. And simply, every discovery has characteristics or physical qualities; mass, charge, spin, diameter, etc. All physical properties. And when you test for all the properties and the tests are positive, then and only then can you say it is a Higgs Boson or a Black Hole. Until then, you are guessing or "Theorizing",(the new word for guessing, used to be a Hypothesis), OR pure Science Fiction. What good does guessing get us? So, can we test everything? No, but we should try at least to draw a comparison to something that is real. There is no "infinite", Theoretical, Singularity or points that can't be defined or measured in Engineering. 3D Space with Time separately. No fourth, fifth, sixth, seventh, eighth, ninth, tenth, eleventh dimensions. Quit wasting time and money on things that can't be proved and therefore are not true. Prove it, put up or shut up! So, for my Universe, An Engineering View.

And a word about calculations. I hear people say they don't have to test because they have the calculations. What??? Math is just a language used to describe something. Some people want to make calculations magical and mystical, like it figures this out for itelf or correct mistakes. Let's do a test to see how smart calculations are: 2 2= What's the problem? This should be an easy calculation. Why isn't it doing that magic. So you see Math will only do what it is told. 2+2, 2-2, 2x2, How can calculations prove anything, except maybe that you are a genius or an idiot! Math is just a language. If I have 2+2=4, I could just as well say in English; two plus two equals four. No difference in the result. Math is just a specialized language that makes it easier to write. If you write 2+2=5, guess what Math didn't correct it. When you Theorize and do the Math, of course it agrees with you, you're the one punching in the numbers! My Calculator and Computer can do some pretty high level calculations, but the output is only as good as my input. Calculations as proof? Show me the calculations to make My Pigs Fly! How does that saying go? Numbers don't lie, but you can lie with numbers, or do calculations.

SUPER THEORY

As I pointed out in the previous Book Vol III, there is a possibility that the Universe is made of only 1 particle. But why could an Engineer Unify the Universe when the best Physicists in History couldn't? Well, for a couple of good reasons:

1. I am not bound by the Theories of Physics. To be a Physicist you would go to school and learn to recite Theories verbatim. As an Engineer I have no such requirement. I can look at individual pieces of a Theory and take the best of the best to Unify the Universe. As an example, if I thought that Quarks could be used to Unify the Universe, I could take that concept only and use it, but ignore the rest of Particle Physics. A Physicist would have Hell to Pay if they tried to disprove Particle Physics. They don't even attempt it. So in other words, Physicists can't Unify the Universe, because they would have to Unify the hundreds of Theories out there. They are sort of trapped by their own inventions.

2. Engineers are very good at figuring out how things work. The mechanics of things, Physicists, not so much. Engineers make things work because we figure out WHY things move and then duplicate it. The Universe can be thought of as a giant machine and something makes it move, Call AN ENGINEER!

So, I have been working on the mechanics of this Unification as

I wrote the book, because the book has pointed me towards it.

And as I pointed out in this book, there is only 1 Force in the

Universe; The Force of Positive and Negative. Now logically

.

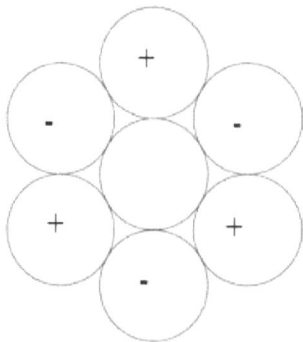

and traditionally this requires two particles. But something just doesn't seem right. So I channeled Einstein, but he was busy finishing Extra Special Relativity. Ok then I prayed to God. That sounds crazy but how could an Engineer, who knows very little about classical Physics, write three books about the Universe, including a Unification Theory, unless someone was giving me the answers. But Whatever; For whatever reason, an answer came to me. Voila, **Super Theory**. It goes like this: In a Star with the Pressures and Temperatures a single uncharged particle exits the core, the RMH-1 particle, God said I could name it! But I do have to mention that it is the real God-Like particle. In the Star these particles become charged, positive rnegative, likely reason would be direction of Spin.3.   Quite naturally some start bonding. Well, apparently around 1836 of these bonds form a neutral sphere, this is the stability limit. Just to recap, 918 negative RMH-1 particles bond with 918 positive -

1 particles. Then, there is room for 1 more positive particle and,

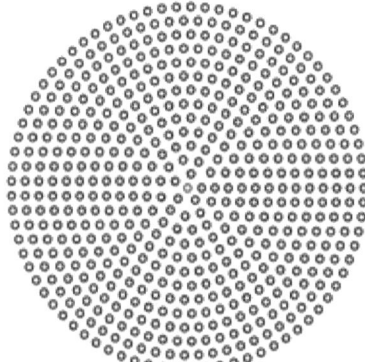 you guessed it, you have a

proton. This explains why a single RMH-1 particle negatively charged -1(an electron) can have the same charge as 1837 RMH-1 particles(a proton), because 918 are negatively charged and 919 are positively charged. Where did I get these numbers from? Simple, if divide the estimated size of an Electron into the estimated size of a Proton, this is what we get. So the net charge for this sphere(a proton) is +1 But at this point, I have to be totally honest: because of the stacking of charges, a proton appears to gain a little extra, so the charge works out to be closer to +1.5. And it is this little extra that allows a proton to bond with a Neutron, which surprise, surprise is a Proton(+1.5) that has captured 2 electrons with a -2 Charge. Again, to recap a +1.5 proton captures 2 electrons(-2) to form a -.5 neutron. And this is why a -.5 neutron bonds to a +1.5 proton so tightly (the Strong Force). Electrons at -1 and Protons at +1.5. And the Star begins to make Atoms. Only Hydrogen has this Electron/Proton combination; all other elements have an Electron (-1), Proton (+1.5), Neutron (-.5) combination.

Even though we can get to a one particle Universe, There is some clean up work to tidy this all up that needs to be done. And that would start with energy levels. It appears that in this model, as well as in the standard model, electrons are capable of carrying more or less energy. And to answer that question it is like a big battery with a 1.5 charge or a little battery with a 1.5 charge. Charge remains the same, just the amount of stored energy is different. Just a few more details. So that's it, a one particle Universe. The mind of God. Oh, and before I forget. Particle Physics thinks that there is some particle that gives other particles Gravity. Well this one little particle already has Gravity and since everything in the Universe is made from it, no need for a Gravity giving particle. See the sub-atomic particles disappearing very quickly. An don't misunderstand, it's not that 30 tiny particles can't exist, It is that there is no need for them. Why make them if you don't need them With one particle I have the Electron, the Proton, the Neutron, Have the Strong Force, The Weak Force, Gravity, and the Electromagnetic Force. I can make Atoms and Elements. I can use my one particle to make all three types of Radiation as well as Planets. One particle can create Hydrogen and create Stars. And when people wake up, the particle will make the Ether

GRAVITY

Before I define Gravity, let's look at some facts. Most of Newton's Gravity works just fine, but Einstein was sure that Light could be bent by Gravity, but how? Since the mass in Light was taken out and replaced with a massless Photon, Gravity should have no affect on Light. So Einstein did something that I think was truly amazing; he looked at the mechanics to try and see how Light could bend! This is exactly what Engineers do! We figure out how things work and why! THE MECHANICS! Tell me that his days in the Patent Office didn't have an effect on Einstein. Anyway, Einstein missed by a mile, but at least he tried! Missed by a mile? What do I mean? Einstein is a Physicist God! Well, Einstein came up with the Theory that Spacetime is distorted by mass objects and Light simply follows this distortion and appears to bend. Brilliant, but wrong! Since Space is thought to be mostly empty, a little dust or stray particle, but mostly empty. SO, THE QUESTION IS; HOW CAN EINSTEIN'S GRAVITY, THAT ONLY WORKS ON MASS OBJECTS, DISTORT EMPTY SPACE? Not one Book, Video, Paper, Lecture or one Physicist asks how Gravity can distort empty SpaceTime. Yet all can recite word for word how Einstein's Gravity distorts SpaceTime. And based on this, either Light has mass(Photons don't exist) or Space has mass(an Eather) or both. Missed by a mile Mr. Einstein!

I didn't want to revisit Gravity, because like everyone I don't see an easy solution. But that is just ducking the issue. How could I claim to Unify the Universe without a complete explanation of Gravity. In previous chapters I gave bits and pieces, but not a complete thought. We'll start at the beginning and try to build the case. At the basic level, Gravity is an attractive force. Well. That is a nice statement, but how does that happen, what are the mechanics. This is where Physics falls on its face; They talk about events but never show what it takes to make it happen. If you can't explain what is really going on, just make up something. Let's call it a Gravitons or SpaceTime. Thousands, millions, billions around, but of course one has never been discovered. So what is the real attractive Force? Ok, I will go slow for Gravitonian Physicists. Everyday Suns all over the Universe pump out vast amounts of Electrons, no not Photons no such things, which makes all the Stars positive. Tesla had this same opinion, but of course he was not German so he had little interest in figuring out the Universe. So here is the honest answer. ALL MASS OBJECTS TEND TO HAVE A POSITIVE CHARGE! Why? Because all mass objects tend to spin. I can explain why, but then the Physicists would have nothing to do, so maybe book V. In spinning, simply, centrifugal Force (And this is why it should be included as one of the Fundamental Forces) is generated and Electrons, especially in the outer orbitals, can be stripped away from mass objects, and then can be absorbed by other atoms OR be thought of as Beta radiation OR drift in Space and help form an Ether(bad word but correct concept). Remember the Nucleus is considered to have a tight bond or Strong Nuclear Force and electrons, whizzing around and away from the Nucleus, are thought to be controlled by the Weak Force. No wonder Physicists can't Unify the Universe, they can't keep their Forces straight! So Here, for free, are the real Four Fundamental Forces of the Universe:

1. Attraction Force.
2. Repulsion Force.
3. Centrifugal Force.
4. Momentum (Stored) Force.

There are, of course, other candidates like my favorite Vibration(Frequency), but are they truly Forces or simply resultants to an applied Force. So real simple, as is most of the Universe. Hey, I bet we could Unify the Universe!

Somehow, this chapter on Gravity has gone sort of sideways, so let's explain Gravity now that we have laid the proper groundwork. Gravity, as is all traditional Forces, is just the Attraction of objects because of their dissimilar charges. Just like Static Electricity, Gravity is caused by an abundance of, or lack thereof, electrons. As I started this chapter, all mass objects tend to be Positively charged as the Nucleus of an Atom is less affected by the spin or Centrifugal Force as the outer electrons, where Centrifugal Force is greater (Duh). Electrons are lost more easily and mass objects tend to be Positive. Ok, here is the tricky part, the bigger the object(mass) the more Positively charged it becomes, Think of Suns with tremendous heat and high Centrifugal Force and the electrons that must be emitted. So here is the slightly tricky part; a higher charged mass is looking for equilibrium, in other words a neutral charge. Yes it is looking for Negatively charged object because that object would have more electrons to give, but in reality it needs electrons, if it is +5 and you are +1 it will try to get some of your electrons. In the broad sense, Positives can attract. This is only true if the charges are dissimilar, if the charges are equal they will repel. A little tricky, but nothing special, just a quest for equilibrium. Remember in Nature, the biggest mass or Force Rules. That is why our Sun can hold the planets in orbit, even though they may all have positive charges. The Sun has a bigger charge. BUT, BUT isn't the Earth a Neutral charge. The Sun wouldn't care if the Earth were -1, 0, or +1 as long as the Sun had the bigger charge the Sun would attract the Earth. But to answer your question; being a mass object it carries a positive charge, but our atmosphere carries a negative charge so it masks the positive charge at the surface and at the surface it appears neutral. But the Earth is extremely more massive than the atmosphere. To the Sun and the Universe, the Earth is Positive, the surface is just masked.

Ok, Just to recap before we forge ahead. Mass objects tend to lose electrons and thereby becoming positively charged. This is because electrons exist at the outer limits of atoms and thereby are the easiest to dislodge or "strip". The larger the object the more electrons it will lose. Thereby, a Sun will be more positive than a Planet. A Planet more positive than a Moon. A Moon more positive than an Asteroid. An Asteroid more positive than a Bowling Ball! Dissimilar amounts of charge attracts objects, not necessarily just positive and negative. A +3 Sun and a +1 Earth is the same strength as +1 Earth and -1 Atmosphere, a 2 differential. Enough recapping let's get to the good stuff!:

This explains Attraction, but what about Repulsion? Repulsion is simply mass objects that have approximately the same charge. The equal lines of Force will separate the objects to a certain distance, and that distance will be determined by the strength of the fields. This is a possibility for Expansion, Oops wrong Chapter, but you can see how things are related. It's called Unification!

Now this is really good Stuff, Attraction and Repulsion, but how can this fit into a Unified Universe. Ok, little Einstein's, Here is how it works. We have just explained how mass objects are

Positive. Well, suppose all of the free or lost electrons floating in Space, evenly repelled each other and formed into a sort of matrix (An Ether, I know, dirty word) with a negative charge.

 Let's see, Positive Mass Objects in Negative Space; Wouldn't that cause an area of distortion around every Mass Object as the opposing charges attracted each other? Hey, isn't this Einstein's Theory of Gravity? Yes, and see how easily other things fall into place. This matrix of electrons can conduct Light and Gravity. And actually limit the speed at which they travel. And that one little particle, an electron, can make the whole Universe. If it spins one direction it can be a negative charge, if it spins opposite it will be a positive charge. These little particles can then bond and make Protons or Neutrons and anything in the Universe. One little particle. But take my word for it, there are NO photons, they simply don't fit. See how easy Space is when you look at the mechanics. An Engineer built this Universe from the ground up. Think about this; I know very little about Physics. I am not a mathematician. I am an Engineer and just by understanding how things go together and move, I can give a

Theory for just about everything in the Universe and more importantly how it is related, in fact, a Unified Universe. You can't have some people in isolated groups all working on different things and expect the results to be consistent and communized. As I said before, Each group has an area of expertise, and a Theory to support it. They have their hands full just explaining their own little world, let alone trying to incorporate some other expertise. That is the advantage I, as an engineer, have. I can start with a common thread and then build on it to explain each area of expertise. No overriding Theory, proposed by some All Knowing Physicist, sorry that was a little harsh, let's just say Nobel Prize winning Physicist that worked in his own little lab, on his own little project. No one ever said, oh by the way, Your Theory must be compatible with the other Theories out there. Not the Nobel Prize Committee, not the National Academy of Physics, no one. No oversite people. If you can get funding, go for it! It is almost better that I am not a Physicist, I have No preconceived ideas of how the Universe works. I have but one directive, everything should work together, simply and efficiently. Ground up approach vs Physics Top down confusion. An Engineering View.

3-D SPACE

People, People, People I am not getting Thru so I will try this one more time. If you ask a Physicist how big the Universe is, the answer may be billions of lightyears across. But what they conveniently leave out is that the Universe is also billions or some number of lightyears thick; 3D, length, height, and width. Why is this important? Because concepts or "Theories" look a lot different if we think 3D. This picture supposedly shows a planet's Gravity distorting Space:

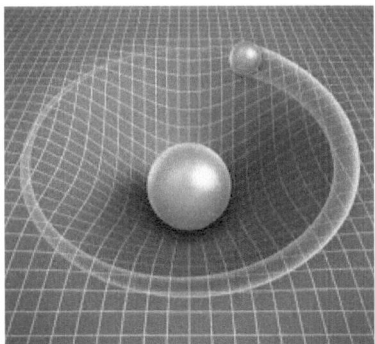

But this is 2D Space not 3D.

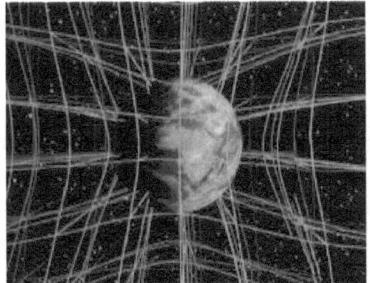

This is closer to 3D distortion. And the point is, the concept changes a little to make 3D work. So, what do I mean? Let's take a look at one of my favorite examples of how things can go so wrong. The Einstein-Rosen Bridge

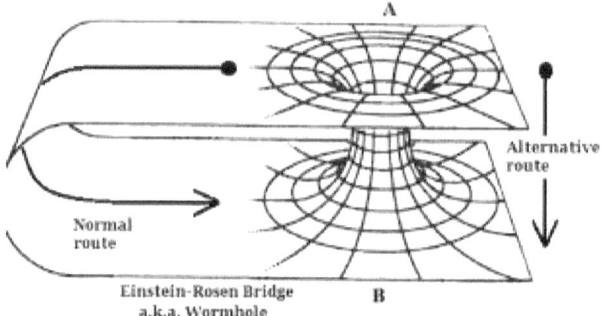

Neil DeGrasse Tyson has a video out where he explains this concept of folding Space and joining the two halves with a Worm Hole, thereby providing a shortcut for Space travel. So please explain to me, Mr. Tyson, how Space, Billions of lightyears thick, is going to get folded! The enery needed would be infinite, AND since Space is billions of lightyears thick, folding it in two would make it 2 Billions of lightyears thick? And you are connecting 2 Billions of lightyears thick with a Worm Hole? Mr Einstein and Mr. Tyson, I just don't think you actually thought this thru very well. The Universe is 3D. BUT, BUT, I must misunderstand. They are only proposing taking a slice of 3D Space and folding it. Oh sorry, I didn't know you had a Space Slicing Machine that could slice a section of 3D Space hundreds of thousands miles, if not lightyears long and

fold it inside The 3d FABRIC of Space. How could I have
missed that? WANT ONE MORE? OKAY …..

 These are Gravitational
waves shown floating on
the 2D ocean of Space.
Who can tell us what is
wrong with this picture?

Yes, it is 2D not 3D. Well that was easy, but what else?
Surprisingly, and I say surprisingly because Einstein did not
mention this, Gravitational waves are not a 2D event. Think of
an explosion or even two mass objects coming together, the
waves will be generated spherically not 2D. No great problem
with 2D pics to help explain, but at some point, as an Engineer, I
would like to see the whole story. And to finish this chapter on
3D space, Let's show SpaceTime for what it is. Even though
the concept of SpaceTime is wrong, I will stick to the Concept of
Space for this chapter. How nice it must be to say something
exists, but never have to define the physical properties that make
it so. Such is SpaceTime. How can you say SpaceTime can
distort or curve or even be flat if you don't show the mecahism
by which it can do that. And without properties, it can't be
tested for. Nice to just have to say something exists. So since
Einstein was not big on the Physical Universe, as an Engineer I
will try to interpret his Theory and show the mechanics and what

it really means in a Physical 3D Universe. Einstein believed that

Space was naturally flat and evenly space unless distorted by something. So based on those words, here is a picture. Einstein

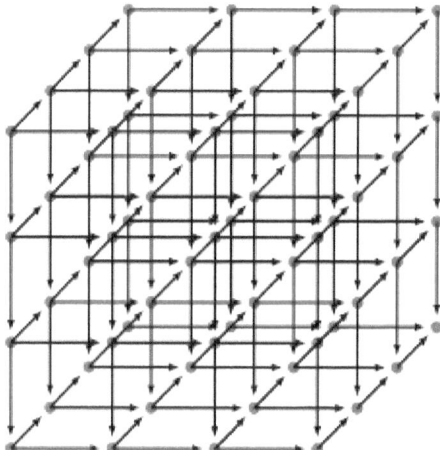

and practically every Physicist love to talk about the top layer, but it's a fact, Space is 3D! So let's show it correctly. And if we

do, Now we can apply Physical properties to produce this
concept. In Space, if every joint were a particle with equal

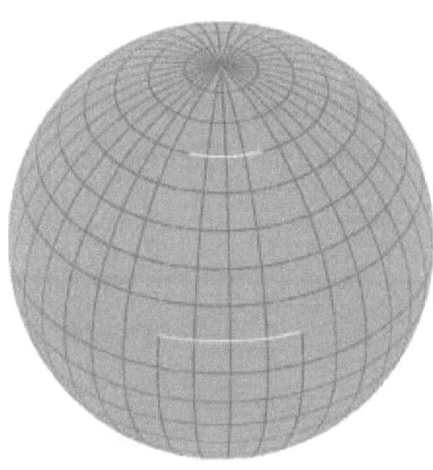

charge. The repulsive
force could cause a matrix
of this shape. We don't
know if the charge is
positive or negative, or the
size of the charge, but we
have something that we
can test and meets
Einstein's description. So,
for the poor Physicists

who can't seem to visualize 3D. Imagine this picture Billions of
lightyears long and Billions of lightyears wide and Billions of
lightyears thick. Now close your eyes and say to yourself.
Where is all that damn Dark Energy? And where is all that Dark
matter? Now open your eyes, look at the picture and say. Oh,
there it is! Right in front of my face! For Einstein this matrix
was his "normal state". And building on that Physical property,
all joints would attempt to reach equilibrium with each other,
evenly spaced. So no curvature naturally. Not that curvature
can't happen, but it will take a force to make it happen. Good so
far? Now for a little more to the puzzle. The question as to
whether Einstein's Universe is Open, Closed, or Flat has just
been answered; We have agreed the matrix of Space is naturally

flat unless distorted by a Force. Now try to imagine how much Force it would take to curve a Universe that was approx. Billions of lightyears in Height, Length, and Width. There simply isn't that much energy in the Universe! This shape of the Universe would not exist because of the energy required to compress Space at the top and bottom and around the sphere, the most energy required.

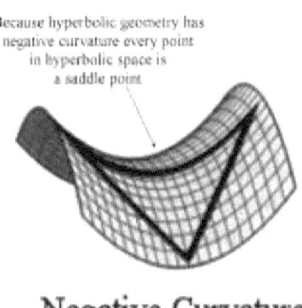

Because hyperbolic geometry has negative curvature every point in hyperbolic space is a saddle point

Negative Curvature

This curvature is not possible either because of the energy required to bend the Universe, but is more possible than a positive curvature. An Engineering View of the Universe. This is easy. Only Flat is possible!

Now the only task is to test for the Charge at the joints; + or − and how much. This would be so much easier if Physicists could see 3D and knew Mechanics. Now what in the Universe is abundantly available that can carry a charge? Oh, I know, how about Electrons. Makes sense, Mass objects positive, Space negative. I those massive positive objects would distort a negative Space, Perfect! So let's see if we can make Physicists brains Explode! Just kidding. But now that we have shown that 3D Space is made of something and therefore can be distorted by Gravity, Let's get crazy. Since the 3D fabric of Space has charged particles at the corners, I bet this matrix could act as one Big conductor. Electromagnetic waves could actually follow along it in Space. Hey, aren't Light and Gravity really just electromagnetic waves? In fact, almost everything that travels in Space is an electromagnetic wave. From Gamma rays to Radio waves. This matrix of Space is really coming in handy. Let's give it a name. I know, how about an Ether? NO, Einstein threw that out and replace it with....SpaceTime! Which looks remarkably like the Ether. Why didn't I think of that?? But See how simply this all ties together if we just start thinking in 3D. Ok, one last thing that people ask and are interested in 3D Space; Are Warp drives possible. Well Yes, but not exactly as shown. So, as usual, Warp drives are not shown in 3D Space, but sort of moving on top of Space and surfing the waves it creates. Close, but not quite. A Warp Craft would be in Space surrounded by 3D Space. So, from a technical standpoint a Warp Engine could still Warp 3D Space but will need a whole lot more power. And I bet you are thinking, what makes him so smart and how does he know? Well, first I am not particularly smart, I just pay attention. And the fact is hundreds of underwater tests have been ran exploring producing(shock) waves in a 3D medium (water). And one basic result was that it took significantly more power to produce waves under water than on top. Extrapolating that to Warp drives means huge amounts of power needed. But another thing to may be limiting is that the fabric of Space may limit Warp Crafts to the Speed of Light just like Light. But if we could achieve Light Speed, I think we would take it. My bet is something less, but certainly possible. On to the 3D World!

Well you thought you would get away, but see, that's the problem with Unifying the Universe; If you are Theorizing about one thing you are really Theorizing about all things! So here is what I found in my Research; Astronomers believe that some Galaxies are moving at much faster than the Speed of Light. And since Astronomers have been pretty precise in their Observations, there is little reason to doubt it. I have been writing this in my books and this proves it. "The Ether or SpaceTime or Fabric of Space or simply Whatever, is the conductor of Electromagnetic Waves and can only transfer Energy so fast, and that happens to be The Speed of Light. "MASS OBJECTS HAVE NO SUCH RESTICTION AND THEREFORE CAN MOVE FASTER THAN THE SPEED OF LIGHT. THE AMOUNT OF THRUST IS THE ONLY LIMITING FACTOR." I now have proof that this is possible. And again, it's all tied together, So if one is true then the other must be true. If mass objects are travelling faster than Light,

Then electromagnetic waves are being limited by an Ether.

Well, that was easy.

Oh God, it's all related isn't? Okay, a little more, but sorry if I cuss. So, Mr Einstein, what about this thing that mass objects can't go faster than light, because infinite Speed equals infinite Mass. What a load of Crapolla. We have whole Galaxies moving faster than Light. Yes, you know what's next, Mr. Einstein, Time Dilation. More Crapolla. Is time slowing down for those Galaxies moving faster than Light, maybe they will reach a Time when they never formed! And by the way, if your clock does slow down and doesn't match Earth's Master Clock, FIX YOUR FRICKIN CLOCK! Sorry God.

THE ETHER

This is the most important chapter in all my books, because if this is proved true, so many questions will be answered and Unifying the Universe becomes much easier. Pay attention! Ok, we have the Unified Forces headed in the right direction, but there are so many other things that are not correct. People ask me after the 4 Forces, what would you correct next? Without a doubt anything that affects the movement of the Universe. And the biggest would be the The Ether. So many movements in the Universe can be explained by an Ether. There has been debate for Centuries about the existence of an Ether with Einstein eventually trashing it in favor of SpaceTime. Now this story is in my first book so I will not be lengthy, but I will be much clearer. The evidence now supports an Ether although we come about it inadvertently. Quite simply an Ether is a medium in Space that can carry a Force or energy. How could this be? Well there are several Theories presented, but I will give you the best (mine). Again, the Ether is just a matrix of free electrons that have formed a sort of lattice in Space. (Repulsion of like charges supports this). Where did the electrons come from? Everyday millions of Suns are spewing out millions of particles and this has been going on for Billions of years. With sufficient quantities of electrons it is not a stretch to see a matrix forming. But how? Ever hear of a thing called Entropy. Electrons initially would be centered over their Star, but because they all carry the same charge. They would repel each other and begin spreading out over billions of years and reach an equilibrium or matrix of Electrons. And it would be easy to imagine this matrix in the whole Universe. Voila, an Ether! Okay, Okay, maybe it's possible, but what is the point? What is the big deal or what can it be used for?? Well Let's talk about the proof first. Einstein Theorized that a massive object would have sufficient Gravity to warp Space around it. And if a beam of light came close enough it would follow this warped Space around the large object. To an observer the Light appears to bend. Brilliant! But there is one problem, Einstein trashed an Ether and the prevailing Theory of Space is that it is mostly empty except for occasional dust and a stray particle. So, if Space is empty, how can Gravity deform empty Space and how can light bend? **Oh, Oh I have an answer;** An Ether. So this is our proof of an Ether, thanks to Einstein. And by the way, when Einstein published SpaceTime he still did not define what Space is made of, but Physicists just keep working with it as though it is a real entity.

Still not convinced? How about this one? There has been a lot of excitement lately about Warp drives that can warp space to

travel on. Same question: If space is empty what is warped?
Let's try an Ether.

One More? Tough Crowd. What about the Einstein-Rosen
Bridge. You know, fold Space in two, connect the halves with a
Worm Hole and create a shortcut through Space. But how do we
fold nothing? Must be something in Space, maybe an Ether.
Ok, those examples were nice, you say, but nothing too Earth
shattering. Why would we really need an Ether? Maybe it could
help explain Dark Energy and Dark matter with its mass and
energy. And if the Ether is expanding because of the repulsion
of the free Electrons, maybe it's powering the expansion of the
Universe. Now I know you are a Tough Crowd, so how about
this one; We could get rid of that silly concept that "**Nothing
Can Go Faster Than The Speed of Light**". BUT IT CAN!
Why? Because Think of Electricity traveling thru a wire. It
travels very fast, not near the Speed of Light. Why? Because the
Electrons in the wire can only transfer the energy so fast as it
passes it down the wire. A Fact. Now suppose that the Ether
conducted Light thru the Universe. Same rules apply, the Ether
can only pass on the Light Energy so fast and that happens to be
the Speed of Light. **BUT,** a mass object has no such restrictions.
In other words, if we can build a rocket engine powerful enough
to go faster than light, there is nothing preventing it from **going
faster than Light.**

So the conclusion to this discussion on an Ether should be; We have all the particles we need, why search for more. Take the money being spend on particle research and colliders and nail this Ether thing down. Potential to go faster than Light will have way more benefits. Think of exploring the Universe.

NOTE: AS I finish this Chapter, I need to point out some inconsistencies that I find in Physics. First, Physicists believe that Photons exist but have no mass, but then the "infinite speed equals infinite mass" rule does not apply. So no mass should produce infinite Speed. What could limit a massless particle to 186,000 miles per second? (I say An Ether). Further, all mass objects in the Universe have energy and the amount of energy limits what they can do, so how can a massless energyless particle even move! All other particles in the Universe have mass and therefore convertible energy. A massless particle in Light? Walk out in the Sunlight and feel the amount of energy being distributed by the Light. So considering that Einstein originally thought there was a mass particle (electron) in light. This massless particle thing (Photons) is utter nonsense. Massless Light should have infinite Speed and not carry energy. So I know that you are an extremely tough crowd, I will give more proof there is mass in Light. Physicists theorize that a massless particle can only go the speed of Light. In fact a massless particle moves at the speed of light as soon as it's created. That is called instant acceleration, because with no mas to accelerate, yes that makes sense. But then Physicists claim that Light is not instantaneous because whatever is emitting the Light needs to build up an energy release. Well you can't have it both ways Physicists, if a massless particle is always going the Speed of Light, then Light is instant on, OR if Light is not instantaneous, then there is a mass particle in Light. And based on the fact that Light has a Speed Limit, THAT CAN ONLY MEAN that Light has Mass OR there is an Ether, because a massless particle should have no Speed restriction.. Something is obviously limiting the Speed of Light. Which is it Physicists, Mass or Ether? OR my personal favorite, BOTH!

When I write these books I do research and try to be as thorough as I can. I look at multiple sources and try to piece an opinion together rather than take one source. But because there isn't much test data to review, it is difficult to find the truth, but even with that, some things just don't make sense to me, like a massless particle.

So let me finish this chapter by why Einstein's Gravity doesn't

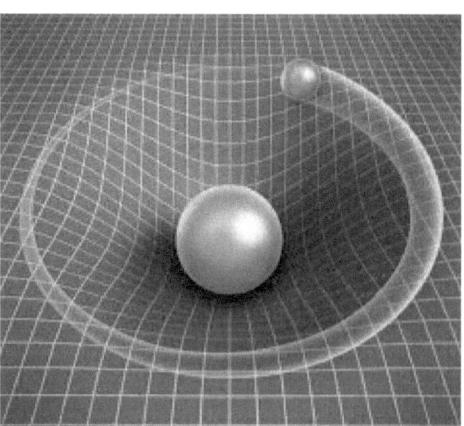

work and Newton's does;

Aside from the fact that this fabric of Space is not defined by Einstein, look at the shape of the orbit of the ball. Yes, it is circular. This is because Eistein's Gravity distorts equally around the sphere and the smaller ball follows that distortion. Einstein's Gravity would only allow for circular orbits, but guess 1 planet that doesn't have a circular orbit but a slightly ellipse orbit. Yes, that's right, The Earth! According to Einstein's Gravity, we can't exist. Take a look;

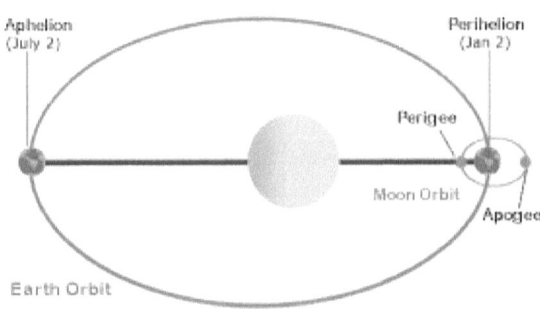

This is the orbit of the Earth, Try to fit this into Einstein's distortion above.

Newton's Gravity easily allows for these elliptical orbits because it does not rely on the distortion of Space only Gravitation bond between objects. This means that if a planets speeds up in the straighter sections it will have to make wider turns at the ends. That will make an elliptical orbit. No so with Einstein's distortion Theory, the distortion will make only circular orbits. So what is really happening in the Universe? Oh my God, there are elliptical orbits everywhere! Wonder which Theory is right? Of course this chapter is about An Ether, and the point of all this discussion. And in the end, what is Occam' Razor or the simplest answer. Here are the answers that an Ether could help solve; Dark Matter, Dark Energy, The missing Mass of the Universe, Why Light travels at the Speed it does, why the Universe is Expanding, why Warp drives could actually work, Why Wormholes could work, what the fabric of Space is made of, and probably a few more I missed. Think this is just a little important?

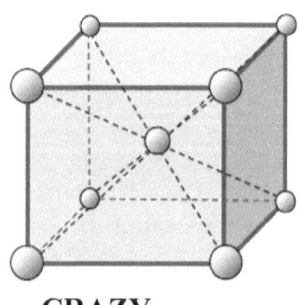

CRAZY

There are some things that are so wrong I can't understand how anyone can believe them. Such is the case of SpaceTime. But let's just take the part where Space gets distorted by Gravity and can bend Light. Besides the fact that NO ONE has explained how empty Space can distort, we will try to understand the mechanics of how this can happen.

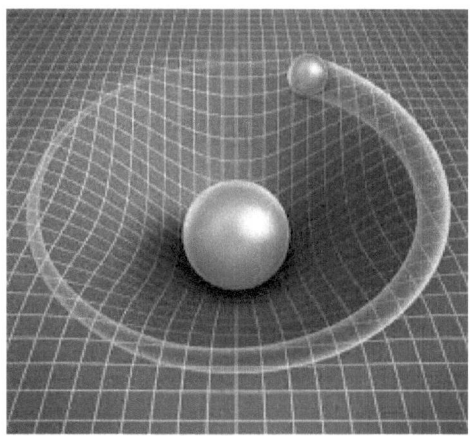

Looking at the picture from above, this is a Sun distorting Space and a planet orbiting the Sun. But not really, as Einstein said the planet is simply following the distortion. So please note that the Sun and Planet are in the distortion zone. Now try to picture a light beam approaching the distortion from somewhere in Space. It is outside the distortion approaching. What happens? Simply put, it would hit the distortion and be deflected AWAY from the Sun. If the distortion is so strong that a massive planet cannot get out, how is a massless lightbeam going to get in? Even though the distortion caused by a spherical Sun should also be spherical, let's just say that for some crazy reason light does get in. Well it's strong enough to break through the distortion to get in, why is it not strong enough to be unaffected by the curvature of the distortion. And since it is affected by the distortion and is curved or bent, why isn't trapped and just continue following the distortion zone. Since this is a Crazy story, I know, I know. As this massless particle begins following the curvature it picks up centrifugal force and gets thrown free of the distortion! A massless particle that was just assigned momentum, breaking a law of Physics, Is now being given centrifugal Force, just to prove Einstein right. This is by far the Craziest story I have heard in Physics. And believe me, it takes some doing to top Particle Physics or the Big Bang people. And I know what you are thinking, What about String Theory!? First of all It's not a Theory, theories require proof. Second, it has eleven dimensions so it's in a class by itself.

YES WE CAN GO FASTER THAN LIGHT

I just finished a rant on the Ether in the last Chapter. Let me say this; The key word in Electro-magnetic Waves is Electro. Just like Electricity, they both need a conductor. In the case of Electricity it usually is a Copper wire, but can arc across different mediums. In the case of Electromagnetic Waves it is an Ether. Just like Electricity and Copper wire, it is the conductor that limits the speed of energy transfer in Light(Electromagnetic) Waves. No so with Mass objects. Mass objects have nothing to limit their speed. This why when you shine a light from a moving train, the Light still will only go the speed of Light. But if you run down the isle on a moving train, your speed will be the speed of the train plus the speed you run. For Mass objects, speed is additive. Do this experiment; Have someone drive a car down your street 40 MPH. Take a Tennis ball and right when they are even with you throw the Tennis ball down the Street alongside the car. You will see you have a hard time throwing the Tennis ball 40 MPH. Next ride in the car driving 40 MPH and at some point throw the ball out the window towards the front of the car. You will find that you have no trouble throwing the ball past the front of car. That is because for mass objects speed is additive; speed of car your riding in PLUS the speed of your throw. Simple, right? Well Physicists don't seem to get it.

Everyday the Earth is orbiting the Sun at 67,000 MPH. That is really how fast you are moving, plus or minus the rotation of the Earth. Patience, I will get to how to go faster than Light. So when the Physicists see Galaxies are moving away from us faster than Light, they claim nothing is moving faster than Light, because the Universe is expanding and the Galaxy is moving. This is so wrong! Why? The formula for speed is distance/time. Therefore, if you travel 60 miles in 1 hour your speed is 60 MPH. There are no qualifiers in the formula like; unless you have the wind at your back, or unless you are wearing rollers skates or riding in a Jet plane OR especially, If the Universe is expanding. Speed is speed, period. Don't try to defend a wrong Theory by making things up. So now that we have proved speed is additive and things are moving Faster than Light, Here is how we can do it. Now there are already proposals to use Nuclear Bombs to power a Starship. This happens to be THE SPEED IS ADDITIVE Theory in concept. You set off an explosion behind a Starship and the blast will push the ship forward and accelerate it to a speed, But you do that every minute or so at the speed will become additive. As an example, if the first explosion can power you to 20,000 MPH, 10 in a row could take you to 200,000 MPH. Additive speed. Now these numbers are not exact, but the concept is correct. But what is left out is the end to this sequence. Since speed is additive, the Theory would be that if you kept on going eventually you would going faster than

Light. No Expansion needed. Now, of course, you have no way to stop, But you could turn around and reverse the process to slow down and, by the way, your Ship would be loaded with nothing but Nuclear Bombs. But the question is not whether it can be done, but How? The real point of this was not to give a working blueprint, but to show, for Mass objects, speed is additive and there is no speed limit. Electromagnetic Waves do have a speed limit and it's the Ether.

THE BIG BANG

So I reaseached and researched but I could not find what I needed to explain how this could happen. I was searching for the mechanics of how people writing about the Big Bang made the walk from all the mass in the Universe down a ball the size of a marble or less. Of course there was no walk, just some references to the Laws of Physics breaking down and that's how a marble came to be. So I forged on, hoping some Big Bang Theory could explain how exactly the whole Universe can come from a marble. No Luck, just assurances that it can. So I gave up and was going to write about something else. Maybe a children's Christmas book. A few days later, I came in from outside and my computer was on and my book was opened to the chapter on the Big Bang. A voice came into my head and said, "I see you haven't made much progress on writing about the Big Bang. I was hoping that you could understand it, because I don't know what they are talking about either. Those Physicists are pretty funny aren't they!" Yes, I thought, but God was laughing. "I'll tell you what I'll do, I'll get you started and then you do your Engineering View thing from there. That Engineering View of the Universe thing is quite the gimmick for getting peoples attention. So give it another shot and see. Oh, and by the way, Physicists are good people they just make laugh sometimes" Thanks God, I said and began writing. I won't bore you with the details, but it goes like this – Nowhere in the Universe do we see compression levels high enough to shrink the Universe down to the size of a Galaxy, let alone the size of a marble. Not in a Black Hole, a Neutron Star, Nothing! The Universe shrunk by compression as far as it could, possibly to a small Galaxy size, then the energy build up would have to release. We see energy releases all through the Universe every day. Gamma rays, radiation, Pulsars, even Suns. We know that compression causes things to heat up, sometimes violently. The most likely think that would have happened is that an explosion took place. It was the mother of all explosions, because it was the whole Universe, and this is important, it was not a center explosion. Because of the distribution of matter, the explosion started more towards one side of the huge Sphere. As a result, 75% of the matter was sent in one direction, slowly expanding. Because it was a Galaxy size Explosion. All Forces were in play, and all Laws of Physics were in play. Nothing too Spectacular. The real question is not how did it happen, but will it happen again? How'd I do God?

BLACK HOLES

I have said this over and over, Black Holes give me a headache. Not because they are anything special, but because the Theory is so Wrong. Look up the definition of "radiation". Alpha, Beta, Gamma, all caused by PARTICLES. If they cannot escape the Gravity of a Black Hole, how can Hawking RADIATION escape? AND if Light can't escape EITHER, wouldn't that indicate Light has a MASS PARTICLE in it, just like RADIATION. At least show some consistency of logic, this is what Physicists do, if you can't solve the problem, just make something up! With this logic, you could predict Clear Holes exist or Blue or Red. Just manipulate the Laws of Nature to what you need! I have a headache. Why do I have to point out basic Laws of Nature. This should be the starting point for discovery. So this is easy, there is no infinity in nature. Infinity is a mathematics term and is right up there with points and theoreticals. Nature has definite lengths, definite sizes, and definite amounts or we should just say Definites. So when you hear things like singularities and infinities or even infinite PI, they are confusing mathematics possibilities with Natural Laws. What does this mean? Well on a basic level, mass cannot be compressed beyond a certain limit. We see compression causing heat. Heat causes energy release or simply explosion. We see it in the Universe frequently, Stars going super Nova or nuclear explosions. So when a Theory suggests that the Universe was compressed to the size of a marble, that breaks the Laws of Nature: An explosion would have happened well before that. Yes these things are mathematically possible, but not in nature. I won't blame this on Physicists necessarily, but they do get carried away sometimes. Black Holes are one of those cases: Infinite density, infinite Gravity? Nothing could escape, not even radiation. Radiation can escape INFINITE Gravity! And massless Photons can't? Did I tell you about Pigs flying? Here is what an Engineer would say about this: You have just proved that Light has a mass particle in it, as a mass particle could not escape infinite Gravity. Therefore, since Light cannot escape, it must have mass in it. But let's go even further and say that the infinite Density or Gravity, You choose, distorts Space and that's the reason Light can't escape, because it just keeps following the distortion around the Black Hole. Distortion does bend Light, you know. BUT, HERE IS THE REAL REASON THAT BLACK HOLES DO NOT EXIST: According to Einstein, remember him, He sort of helped support Black Holes, Gravity distorts Space, and this distortion can even bend Light and Physics claimed it was proven. Well, if Gravity can distort Space and it's proven, then infinite Gravity should distort Space infinitely. DID YOU SEE WHAT I JUST WROTE, INFINITE DISTORTION! We should all be able to look out at the night sky and see distorted starlight everywhere Why don't we, Mr. Einstein and Mr. Hawking. Because BLACK HOLES DON'T EXIST! Why? Because in mathematics you can calculate to a Singularity or a point or a theoretical. But in the real World, the 3D World these things can't happen. The Laws of Nature would have to disappear. Just like the Big Bang people who claim the Universe the size of a marble, you can do that with math, but the 3D world could not go there. See how these infinite Theories just don't make sense. This is also the reason we can't solve for PI, because it is a math problem. In book VOL II, I bring PI into the real world and give you the easy solution. When you hear the words infinite or irrational or Theoretical, RUN. They are talking about Mathematics, not the Universe!

I want to make this simple and perfectly clear. An "Event Horizon" is just a point where gravity is so strong that nothing can escape. If the Gravity is infinite, the Event Horizon is infinite. The whole Universe would be the limits of the Event Horizon! Listen to the words, not Significant, not Very High, but Infinite! Infinite means never ending. You can achieve a Theoretical Point or Singularity in math, but not in Cleveland! And Pigs can't fly! As Theorized and defined, Black Holes cannot exist. An Engineering View. So, Ok, As much as I hate writing about Black Holes, you deserve the Real picture, the mechanics. Just like the Solar System, a rotating Galaxy should have a huge mass in the center. There are 3 options that could have happened:

1. There was a massive Star at the center of the Galaxy holding all smaller Stars in orbit. It's odd that we don't see that much.
2. The massive Starr collapsed into a Black Hole and still holds all Stars in their orbits.
3. There never was anything at the center and the center point was just determined by the amount of mass and distance from the center of rotation that was determined when the Galaxy was throw off by the Big Bang. Kind of like throwing a Frisbee with no center. It will still rotate around a Theoretical center point.

Let's look one at a time. 1. A massive star exactly at the center point would be highly unlikely. Certainly a Star could have a chance of being at the center, but would it be massive?

2. Equally or more improbable as #1 as the collapse would have to happen too. Also, Massive Stars don't usually turn into Black Holes and this Star would very Massive.

3. This seems like the best option. A slow spinning galaxy would not generate much centrifugal Force and this would allow the inner Stars to hold the outer Stars in place through Gravity. The inner Stars acting as an inner Hub. With no spinning Mass

at the center, this Galaxy would slow down a little faster, but what is time in Space.

So, there it is 3 scenarios for Galaxies and Black Holes don't seem very likely for obvious reasons. Add to that Black Holes would have a hard time getting started at the compression ratios required. Slim to none. So let's look at a Black Hole:

At first I thought

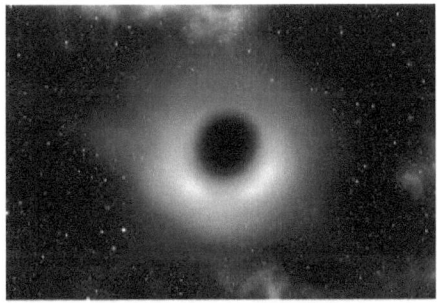

this was a fake, because the Event Horizon of a Black Hole is Spherical, The Fire should be Spherically around the Black Hole, meaning this picture should look like one big ball of fire. In other words, this looks like a slice made through the equator of the Black Hole. Second, If a Black Hole had infinite Gravity, shouldn't the Event Horizon be a much, much bigger diameter? Don't know the scale, but I vote fake! And speaking of fake, wouldn't it be funny if one day we found out that giant Planets existed that were either made of or coated with Carbon. Let's see, what would that resemble? Could happen, maybe the Remains of a burned out Star.

PHOTONS DON'T REALLY EXIST

I am going to give this one more shot. It is really important. In Unifying the Universe, it is important to sort out the things that don't fit. If everything else in the Universe works one way, then how can one thing not work like the others. Or, in Engineering terms, One of these things is not like the others! Let's make this simple so even Einstein could get it. Actually, Einstein did get it, it was the rest of Physics that didn't and still doesn't get it. Get what? Einstein believed there was a particle (electron) in Light. Doesn't sound too radical, but there was a problem; Einstein gave Physics $E=mc^2$. And as speed increases, mass increases. But that means at the Speed of Light, the mass in Light would become infinite. Oh no, what to do? Well, the Physicists found the solution or Theorized! Just take the mass out of Light and the World will be alright again! No experimenting, no testing, just Theory. And Yes, the Photon was born. This was the name given to this massless particle. BUT here's the problem with making things up and not doing Science (Particle Physics), the Photon needs mass characteristics. The story will just keep getting bigger. As in Momentum=mass x velocity. No mass, no momentum, No problem, we'll just give it momentum. But it must be able to carry energy too, again no problem, we'll just say it does. But, as with all made up stories, eventually the Truth comes out and the story collapses. Such is the story of the Photon. What the story-tellers didn't realize at the time, was that Light was an electromagnetic wave, but not just Light, but a whole series of electromagnetic waves, from radio waves to Gamma waves, they ALL got Photons. That's OK, we'll just say Photons are in them too. Story is getting Bigger. Maybe ready to collapse! Guess what was discovered? RADIATION! Alpha, Beta, Gamma. Guess how it travels, as electromagnetic waves! Guess what it has in it? You guess it! Honest to God, real, testable, physical particles! Particles that carry energy and can cause damage to Humans. Now, Let's see, Travels in electromagnetic waves and carries energy. Sounds just like a Light wave. One of these things is not like the others, one of these things just doesn't belong!

Now, I am not a smart man, but if 3 electromagnetic waves have particles in them, shouldn't they all? Tested Truth vs Made up particle? Oh, did we say all, we meant just Light, they will say. Right and Pigs can Fly! And massless particles do have momentum. A Good Story, But a Story No Less! And while we are at it, let's correct the wrong that has been done. A massless particle, by definition of momentum, has no momentum. A massless particle cannot carry Energy. And the biggest correction of all: As speed increases, mass does not increase, only momentum. So let's say this slowly and hope it sinks in; There is nothing special about Light. It is at the right frequency

that we can see. Just frequency differences between ALL
electromagnetic waves. If one moves in waves, they all do. If
one carries energy, they all do. And if 3 have particles, they all
do, I THINK WE ARE DONE HERE.

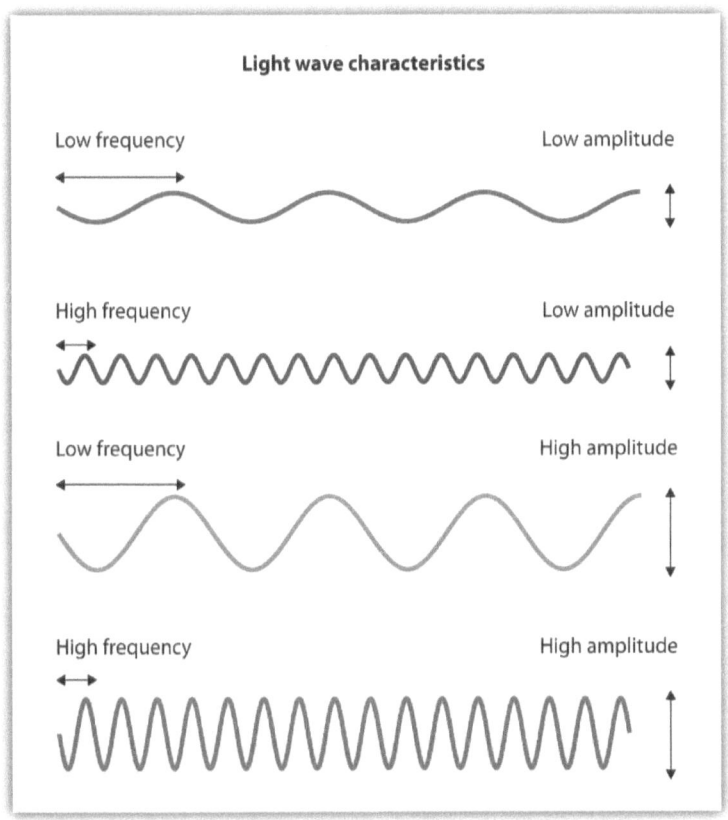

THE HALF OF A HALF OF A HALF THEORY

It's obvious that this is the Theory of no matter how small a piece of matter is, it can always be divided in two. This is absolutely true except for one thing; is it necessary! And this means there is an end point. If I have an Electron and I smash it in two, all I have is 2 pieces of Electrons, not some Quarks and Gluons. Somewhere things have gone terribly wrong. Atoms make the whole Universe. And are made of 3 basic particles; an Electron. A Proton, and a Neutron. With these we can explain everything else; The Strong Force, the Weak Force, Electromagnetism, Gravity, Radiation. Not to mention different phenomenon like Dark Matter, Dark Energy, etc. Can't we just stop and say 2 pieces of Electrons are 2 pieces of Electrons, instead of the Starsky and Hutch particles. We don't need them. They just happen to be an Electron that collided with something and slit in half. Once you are dead, you can't be more dead or once you are insane, you can't be more insane. So stop just before insne! This is also the complex model, once a system is complex it doesn't need more complexity or as I like to think, once you enter the Matrix, you don't need to be more Matixy. Everything in Nature has a limit. Get used to it.

RECAP OF VOLS I, II,III

VOLI

Volume I touches on the big issues in the Universe. I try to stick to the mechanics or Engineering view of events, but also give a simple solution. And along with analyzing each subject, I try to show how it links to other events in the Universe. To be honest, Volume I was the hardest to write as I am not a Physicist and it took hundreds of hours of research to go from the beginning to current day. I already had some opinions and those I had to be filled out. Others I just completely had to start from scratch. So lots of subjects and lots of data, but this book actually lays the groundwork for the Unification of the Universe

Vol II

Vol II was the most fun, as it allowed me to solve for Pi. Just like the Universe, I sometimes think, This is absurd, we can't figure this out? So I have to try. Well, I found an online calculator capable of 100 digits and I began calculating and sure enough I got the same answer everyone else had. So I switched to an Engineering method and I found the mistake and answer. 3.15 take it to the bank. Sorry you'll have to buy the book to get the explanation. But as for the rest of Vol II, there is a little more on the Universe and connecting the dots. I had always thought that it would be a three book set to write about the Universe, but I was just guessing. Considering my lack of Knowledge, even that may be a stretch. But I knew I had a different take on the Universe as an Engineer, so maybe three. After reading about and looking at so many subjects, my Engineering logic began to see commonization and simplification and I decided to tie it altogether in Vol III

VOLIII

This was the Unification of the Universe. I was fairly happy with the way it turned out as the first two books built the foundations and this book tied it altogether. Of course the hardest chapter was Gravity. I was not too pleased with the first two books and Gravity, but then it was never my intention to Unify the Universe. But as I wrote every chapter, I could start seeing the possibility of the Universe coming together. My biggest doubt was Gravity but I didn't see the path. I went back, as I always do, to simpler. And pretty quickly the concept came together. A few experiments with electrical charges and away I went.

SUMMARY

 Occasionally I get asked questions about Space and the Universe and other things that I just don't have an opinion on, because I just haven't spend the Time. Somebody asked if Space is infinite. So I haven't thought about it much, but here is an Engineering View. Look at as much data as you can find, even if it is circumstantial data. On one hand you have people saying that if the Universe is flat, that means it is infinite. But that is silly. If you came across a set of Railroad tracks, they would be fairly flat, but would you say they are infinite. No, so why would you assume Space to be different? And further evidence suggests that one day there will be no Universe, the ultimate Entropy. And if one Universe can disappear, they all can. So simply, no need for infinite Space. So what would create it?

But the real answer is: Who Cares! Can't See it, Can't use it and Ain't gonna be here long enough to think about it. Focus!

Probably the most questions I get asked are about Einstein. Was Einstein a Genius? I believe everyone is a Genius at something. Einstein benefitted by working as a Patent Clerk because he got a feel for how Engineers think. Just as I would benefit from working in a Physics Dept. Einstein wasn't a great Ideas person, as he collaborated frequently, but was excellent at figuring out how things could work in his world, but Ideas were not his strength. In fact one could say that later in his life, he sort of ran out of Ideas. Was he Great? Let's say unique! Newton was my Hero, imagine him with today's technology. But I definitely would want Einstein's opinion. There have been so many Great people.

FORMATION OF THE MOON

So how did the Moon Form? But why stick to just the Moon.
Shouldn't every Celestial body be the same? The short answer is
yes, but there can be deviations. The scale of a creation of a
Moon is not the same as the Creation of a Planet or a Star. Little
amounts of mass can move quicker that big amounts, the Forces
are less, the heat is less. It's easy to see gravity pulling large
masses together, not so much with small objects like a Moon.
Why bring this up? Because I ran across this in researching the
Universe and it is a great example of how discovery should be
done. Some Physical samples of the Earth and Moon and a
Hypothesis on how they could be the same. Computer
simulations to study the mechanics and some more Physical
testing to refine ideas. Studying alternate ideas for what else
could cause the same result. This Methodology closely
resembles the Scientific Method. Yes, Real Science. And just in
case anyone wants to know the answer, there appear to be three.
The Moon could have been a stray object that came close enough
to the Earth's Gravitational field to be captured; reasonable.
And similar to that idea is the idea that stray matter in the Solar
System coalesced thru Gravity and formed the Moon near the
Earth, somewhat less likely. And the third Scenario would be
that some smaller than Earth object collided with the Earth, with

both objects becoming very hot. So much so that they sort of melted together and exchanged some Atomic material. Eventually Earth ejected this off center mass and the Moon was formed. I like this way of discovery. Three possibilities laid out and we get to use our own deductive reasoning to decide which makes more sense to us, as opposed to Particle Physics that tries to push one and only one "Theory" down our throats. With no Physical testing, You could be wrong. Show us some alternatives. Who knows, we could actually help you. So the point is, even if I don't agree with the conclusion, the Science is being executed very well. The Scientific Method originally asked that alternate possibilities be looked at and investigated. How many Theories include alternatives. No complete testing or examination of alternatives, but we Know our Theory is right! Oh, I get it, and Pigs can fly too!

APPENDIX

SOLVING FOR PI

This Chapter was actually published in Book Vol II, but this is one of the more important Concepts in the book. The search for PI was brought about because basically the Greeks did not figure out how to measure a circle. (why measure, we have calculations) Sound familiar, How's those calculations working for Pi? Mille and Zoe, my brain trust (actually Puppies) occasionally look at me quizzically like they don't understand what I'm talking about. Such is the case for PI. Other people have asked me, actually challenged me, about this. I have been putting it off as its not as much fun as The Universe, and math gives me a headache. Ok, so here it is, Occam's Razor, simple answer. THIS IS NOT A MATH PROBLEM. This is more of a Geometry problem. When you use math, it needs to be used to explain what is happening in a physical Universe, In other words, Math is not a Solution, it's an explanation.

$$PI = 3.15$$

Why would an Engineer, writing a book on the Universe, include a Chapter on Pi? One could say that the Universe includes all, so anything is fair game. But the fact is that I notice something wrong; logic, proof, conclusions, etc. and I must respond (the German in me). The truth is this should be so simple I don't know what went wrong for hundreds of years. What do I mean? Pi=circumference/diameter, pencils down, computers off, end of story. For an Engineer this is right up our alley; measure the circumference, measure the diameter, divide, repeat that a few thousand times, then ask for a raise.

Somewhere things went horribly wrong. Engineers did not get involved. Maybe they thought it was so simple even a Physicist, oh sorry, I meant Mathematician, could do it. Whatever, where is the data? Hundreds of years and no data, only calculations. I get it, our precision for making and measuring a perfect circle is only good to 3 or 4 decimal places, but it should still exist even as a directional indication.

What Is Wrong! Let's fix this; even though I prefer physical data, we can break this down. First a little History: The ancient Greeks; Euclid, Archimedes, Pythagoras, et al, gave us most all of the formulas for triangles. Why is this important? Because the quest for Pi has centered around multi-sided polygons and triangles to approximate Pi. It is believed that if a circle can be divided into smaller and smaller triangles(billionths of a degree), the straight line bases of the triangles will at some point match the curved circumference of a circle. But as one University of Michigan student told me; a straight line will never equal a curved line no matter how small the pieces. And this leads to an irrational quest for Pi, That old half of a half of a half Theory. Ok, back to the Greeks and the formulas. For some unknown reason the Greeks included irrational numbers in their formulas. What do I mean? Let's take the Pythagoras Theorem. In a right triangle, the side adjacent squared plus the side opposite squared equals the hypotenuse squared. SO, if side adjacent is 1 and side opposite is 1, the hypotenuse is equal to the square root of 2 which is an irrational number. This is Absolutely wrong! Why? Because these are not just numbers they can be thought of as measurements. Side opposite has an exact measurement of 1 and side adjacent also has an exact measurement of 1 and since the hypotenuse is connected dot to dot from one end of the side adjacent to the end of side opposite it too has a define length. An irrational number is not a length and therefore cannot be a side of a triangle. The Greeks are somewhat wrong in the triangle formulas. Now let's get back to Pi.

Start with the knowns or truths. We know Circumference/Dimeter= Pi. We also know that both the Circumference and the Diameter can be measured from each start point to an end point and therefore are not endless strings, but a fixed length. Wait you say, Pi is not a measurement, it is a ratio and that could be irrational. Nice try Grasshopper but since Diameter can be 1, 10,100 or multiples of 10, in this case Pi would be a fixed Circumference length divided by 1; a fixed number, and not irrational. Yes, 2 through 9 could all still be irrational, but the idea of a fixed ratio for Pi would be disproved. UNLESS, Pi is evenly divisible by 2 through 9 (chart below). Then further, current claims to Pi or any irrational number are WRONG!

I know this hurts, but want more Truth? Okay, a picture:

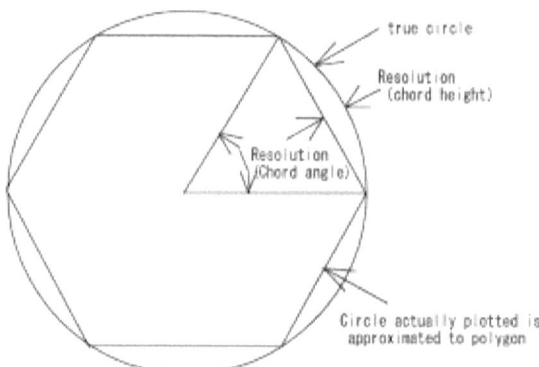

Here is a picture of a multisided Polygon compared to a circle. Notice how much bigger the circumference of the circle looks to the Polygon. But for a 360 faced polygon, the height difference from a straight

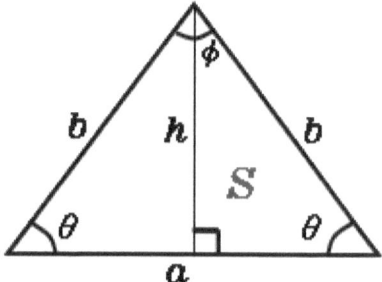

line to a curved circle for any face is approx. .0004! Easy to see why Archimedes thought that 360 degrees made a circle.

But that never seemed right to me; I investigated. With the help of Ke!san Casio Online Calculators, which allow 50 digits for triangles and 102 for general math. I found a small problem with the formulas: The circumference of a polygon can never equal the circular circumference or as my University of Michigan son told me, no matter how small a straight line will never equal a curved line, so why look at straight line solutions? The proof of this is that if we take a 360 degree circle and look at a 1 degree slice or a triangle (one face of a 360 face polygon). This 1 degree slice is an Isosceles triangle with equal sides, with each the length of the radius(b); a bisecting line(h) can be generated but the length will never equal the radius lines, even if the angle at the top is so small that the 3 lines (b-h-b) almost touch each   other.

What   happens is
that if you   divide a
circle into a billion triangles, each height(h) will be short of the radius by a billionth, and a trillion by a trillionth and so on. This not only creates an incorrect irrational condition but is just

WRONG, because we already know a circumference has a fixed length. Even if that length is to a billion of inch, there is an end to a measurement. At this point it should be a perfect circle where all lines(b-h-b) are equal to the radius. So instead of calculating Pi to infinite digits, there should have been an end. Remember, for a circle, the start point and end point are the same point. The measurement is fixed, not an endless series of numbers.

I know what you are thinking; if Pi is wrong, mister smart guy, what should it be AND why? Well, I worked backwards with the calculators to calculate a perfect circle, meaning that all leg lines and bisecting lines MUST be equal to the radius AND Pi MUST be divisible by numbers 1-9 or we will be back in an irrational situation. And Voila, Pi = 3.15!

And as a proof, here is Pi divided by all 9 numbers:

3.15	1	3.15
3.15	2	1.575
3.15	3	1.05
3.15	4	0.7875
3.15	5	0.63
3.15	6	0.525
3.15	7	0.45
3.15	8	0.39375
3.15	9	0.35

Imagine this chart by replacing 3.15 with 3.14159265 with a never ending series of numbers. Why did it never occur to anyone that something is wrong and let's find what's wrong. But NO! Let's just keep calculating Pi with smaller and smaller straight lines, even if the straight line length gets smaller than an atom. And to simplify this explanation, what I have really proposed is a calculator for a circle instead of straight lines trying to achieve a curved line. To be clear about the triangle calculators even with a circle divided into a billion triangles, it shows a 5 radius with a 4.9999999999 bisecting line. Simply put; if the line is correct at the 2 ends but off by a billionth in the middle, it is off 1/3 of the time. And since the circle is divided into a billion triangles, the circle calculation is off 1/3 of a billion times. Easy to see 3.14 becoming 3.15. As I have said this should be corrected to a 5 radius and a 5 bisecting line(height) then you have a perfect circle and 3.15 Pi. Now I have calculated the angles and line lengths to calculate circles, but my Union Contract says I can't tell you. But you have 3.14159 etc., and I give you 3.15, I think even a Physicist, oops did it again, a Mathematician can figure the difference. Pencils down, computers off, where's my raise?

BUT, BUT, BUT, Wait a minute. 3.15 seems so much bigger than 3.14+?? BUT If you subtract 3.14+ from 3.15, I hate math, but you will see the actual difference is approx. 0.008 around the entire circumference or approx.. .0002 per 1 degree! The human eye would not even perceive this change. HEY WAIT A MINUTE, you complain about no physical data(measurements) for circumference, but where is yours Mister Engineer? I have put the measurement procedures at the end of the book in the END NOTES. I would encourage everyone to measure Pi.

So to sum up Pi: Took Euclid, Archimedes, Pythagoras and current Pi to task for flaws, solved for circular Pi and only needed 2 decimal places: Had a good day! Now you have an Engineering answer for Pi

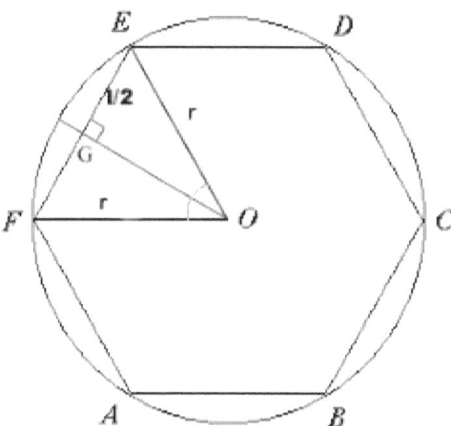

ENTROPY

Yes, another subject that really doesn't affect the Unification of the Universe. But this is Thermal Dynamics and heat does play a big role in the operation of the Universe, so here it is: The second Law of Thermal Dynamics. Well, this is a fine mess

Another subject that Engineering really doesn't get involved in, but I see a flaw, so here I am. Let's just put Entropy aside for a moment and try to understand the Thermal Dynamics part. There are only 2 possibilities: 1. The system is closed and heat cannot escape or 2. The system is open and heat can go into or out of the system freely. Now this is very tricky because the Laws seem to imply they apply to a closed system as an open system could allow for some adding of heat or cold to the system As best that I can understand, Entropy is an offshoot of the second Law of Thermal Dynamics and energy transfer using statistical probability mathematics. Or more simply, over time the (Heat)energy of a system will tend to spread out evenly thru out the system(high Entropy). And during that process the particles(Micro States) may be in any position based on probability. But it refers to the complete system as the Universe

This is seemingly pretty straight forward but the outcome appears to actually be a Hot Mess. Since this is an Engineering View the very first question is:

1. What is the period of Time? A minute, a day, a week, a year or a life time? Engineers live for data and details. Nobody ever bothered to keep Time?

2. As an Engineer, the mechanics of a system is one thing that jumps out right away; What moves?, When?, Why?, How? What force is the spreading the particles out to equilibrium and high Entropy?

3. It is implied that this is a closed system, but unless the thermal shield can isolate the system from the outside surroundings(Air?,Universe?), it's heat will transfer out of the closed system and into the Universe. Therefore leaving behind energy-less particles trapped in a closed container. And then the particles are supposed to form into Micro-states? Do this experiment: get a pan of water, heat it to boiling, then turn off the heat and time how long it takes to cool. At the same time look up how many water molecules in 2 or 3 cups of water. Then divide the time by the number of molecules, this number is roughly the number of Microstates also. This is the approx. time each Microstate has to form and reform. Millions of Microstates would have to form at a blur, and also before the water evaporates into the Universe. Even in a vacuum, heat can be

radiated, just not conducted. The conclusion is there are no closed systems.

And finally, 4. What about Gravity? Even if the heat energy is radiated away from the particles, the particles still have mass and the four fundamental forces of nature are still available, although the strong, weak, and electromagnetic forces don't effect a single particle, Gravity can. And since Gravity, although very weak, has an effect over very long distances, it would be logical to assume that Gravity would begin pulling particles together. And before we dismiss this idea, this is exactly what happens every day on Earth. Earth's gravity causes tons of very fine dust (maybe energy-less particles?) to fall on Earth everyday. Starting with very small particles to build a Universe would take a very, very long time, but who knows, maybe this is the exact way this one started.

And since Engineers look at the mechanics of a system, it would fit nicely that heat is carried by electrons and electrons can carry lower and higher states of energy. Also electrons carry a negative charge that can repel each other and nicely explains the spreading out of particles and Entropy. But no, someone comes up with Micro-states and the random forming of arrangements of particles. How do these arrangements overcome the repulsion force? Just when it looked like this was going to make sense!

2nd Law of Thermal Dynamics and Entropy? Sorry, this Law doesn't seem very well thought thru. And since Engineers invented steam engines before the Laws of thermal Dynamics ever existed we should have just wrote down what we saw and did the Laws ourselves.

HEILMAN'S THEORY OF
PROPORTIONALITY

I introduced this in the Last book Vol III, but only to Highlight the differences in like things. As an example, there are positively charged objects in the Universe; Galaxies, Suns, Planets, Moons, Asteroids, Bowling Balls. It is important to understand the proportional relationship between them, if we are trying to figure out how and why they move (The mechanics). And again the proportion of the Force may or may not be exactly proportional to their size or Mass, etc. So to truly understand The Universe we need to be mindful of Proportions of things and how that actually affects Forces. Gravity is an attractive Force, but there are other Forces that may strengthen or oppose it. The same amount of anything may not yield the same results given the proportionality of the factors. This sounds pretty obvious, but it is very rarely mentioned in solutions to anything. Those are Clean Room solutions. Solutions as though everything is perfect. So I mention this, and give a few examples to show these subtleties can influence a result to either help prove or disprove an assumption. This chart gives an example of have proportionality can work and can be used to arrive at a reasonable answer to complex question or problem. This chart actually shows all the Stars in the Universe. As you can see our Sun is a class G star with eight Planets. So proportionally if a Sun is 1.5 times the mass of our Sun (class F), it should have 1.5 times the Planets as our Sun. And we can look up elsewhere the est. mass of our Planets. So we can est. the mass of our star and planets, and we can proportionally est. the entire mass of the Universe by knowing the number of Stars in our Universe. This chart gives the est. distribution by class. A little math and all we need is some est. of Asteroids and Dust and we have the mass of the Universe! Ok, less Black Holes, but they don't exist.

Main Sequence Stars

Spectral Type:	O	B	A	F	G	K	M
Temperature:	40 000K	20 000K	8500K	6500K	5700K	4500K	3200K
Radius (Sun=1):	10	5	1.7	1.3	1.0	0.8	0.3
Mass (Sun=1):	50	10	2.0	1.5	1.0	0.7	0.2
Luminosity (Sun=1):	100 000	1000	20	4	1.0	0.2	0.01
Lifetime (million yrs):	10	100	1000	3000	10 000	50 000	200 000
Abundance:	0.00001%	0.1%	0.7%	2%	3.5%	8%	80%

Giant Stars
Low mass stars near the end of their lives.

Spectral Type:	Mainly G, K or M
Temperature:	3000 to 10 000K
Radius (Sun=1):	10 to 50
Mass (Sun=1):	1 to 5
Luminosity (Sun=1):	50 to 1000
Lifetime (million yrs):	1000
Abundance:	0.4%

White Dwarfs
Dying remnant of an imploded star.

Spectral Type:	D
Temperature:	Under 80 000K
Radius (Sun=1):	Under 0.01
Mass (Sun=1):	Under 1.4
Luminosity (Sun=1):	Under 0.01
Lifetime (million yrs):	–
Abundance:	5%

Supergiant Stars
High mass stars near the end of their lives.

Spectral Type:	O, B, A, F, G, K or M
Temperature:	4000 to 40 000K
Radius (Sun=1):	30 to 500
Mass (Sun=1):	10 to 70
Luminosity (Sun=1):	30 000 to 1000 000
Lifetime (million yrs):	10
Abundance:	0.0001%

So proportional calculations can exist and provide an excellent est of things in the Universe.

THE SCIENTIFIC METHOD

As I was writing this series of books, I realized that we all seem to be working to slightly different versions of the Scientific Method. Simply put the Scientific Method was developed by the Scientists of earlier times, as a roadmap to discovery. In fact, Aristotle 384-322 BC is considered the inventor of the Scientific Method. But the problem has always been; as it was passed down thru the Ages things were refined especially to keep up with Technical improvements. The early communization began around Newton's time and continued on as more and more Universities sprang up. So, even from the start, the Scientific Method was more of guidelines or Method than a strict set of rules or step. Yes. It was written with general steps but it was realized that all steps may not be needed. I will give the general idea, but this is intended for explanation only:

1. Observe something.
2. Formulate an idea of how or why.
3. Investigate and research. (maybe someone already had the idea or info is available, or a Theory.)
4. Formulate a Hypothesis of the how and why.
5. Experiment to test the how and why and continue researching for additional info or data.

6. Analyze the experiment(s) and research to adjust the Hypothesis and re-experiment if necessary.

7. (This is the most important step and continually gets left out:) Plan to test your Hypothesis: Itemize a list of all the characteristics of your discovery and plan a test for every one. LEAVE NO DOUBT that you have discovered exactly what you hypothesized.

8. Analyze the Test results, including the possibility that other things could may have given you the same results.

9. If necessary adjust the Hypothesis and re-test. If not, Congratulations you now have a (proven) Theory.

Now this a general outline or maybe more correctly an Engineering View. The point is that many Theories could not pass this test. And keep in mind that many Theories are called successful just by proving 1 characteristic. Big Bang is not a Theory, nor String Theory. And almost all of Particle Physics and the Higgs Boson is not Theory. The next thing this guy will be telling me is that there is no Pigs can Fly Theory. OK one more then I have to rest.

THE GRAPH PAPER THEORY

This really isn't a Theory, but just a cool way to dress up your Theory. If your Theory lacks a lot of Physical data, this is an excellent way to fill up a page and look professional doing it. Not saying that Graphs don't have their place, but charts and Graphs and Graphic Presentations seem more like the tools of a salesman than a Scientist. As an example, here are a couple of my favorites: See what I am talking about? Not a lot of data,

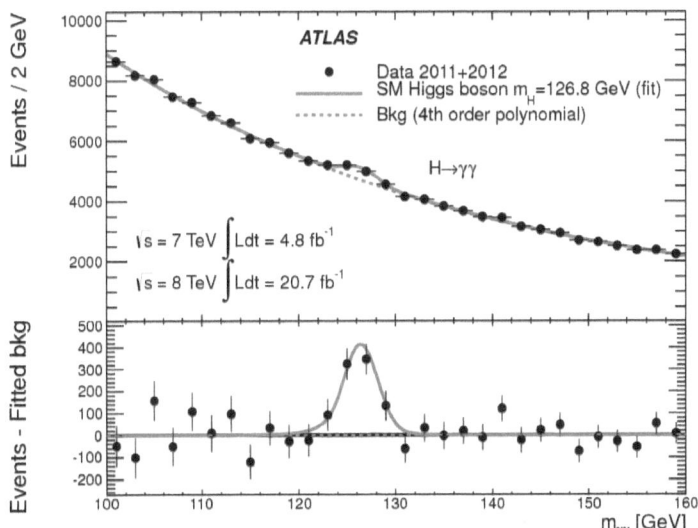

but it is presented well. Keep this in mind if you need something to visually impress.

And another favorite:

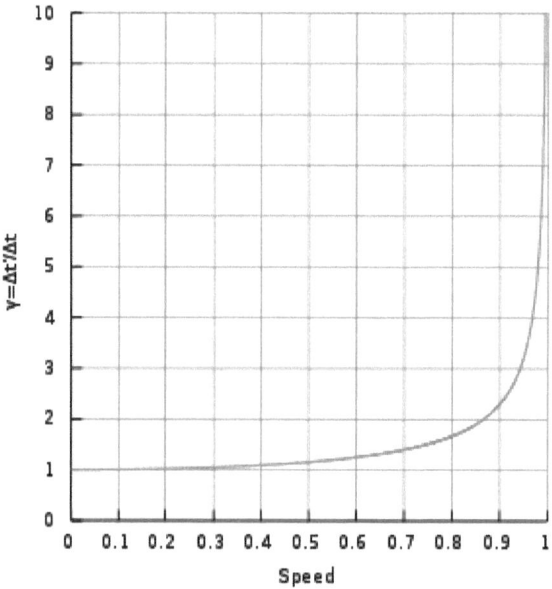

See, so little info or data, but doesn't it just make you say "What is this"? Yes, Time Dilation, but why say it when you have this cool graph. Sometimes you don't even have to explain it if your graph is cool enough. OK, enough fun.

DUST
OR THE MASS OF THE UNIVERSE

This chart was added merely to show that a complete
mass roll up of the Universe was done, including all mass
objects that even included Dust. A spreadsheet was done
in the Book Vol I. The Total, of course, came well short of
Astro Physicists estimates because there were no numbers
for Dark Mass and Dark Energy.

Mass Speadsheet part 2 Dust.

	Dust/year Earth	% of Earth's Mass	((Area(SS) x .98/Earth area) x
	3311225.25	5.543098382884690E-19	Earth Dust=SS Dust) x star
mass x number of solar systems	Dust/year	Star (SS)Mass x Percent	number in Universe SS
9.96E+45		5.52E+27	3.77E+33
1.99E+49		1.10E+31	7.54E+36
2.79E+49		1.55E+31	1.06E+37
5.98E+49		3.31E+31	2.26E+37
6.97E+49		3.86E+31	2.64E+37
1.12E+50		6.18E+31	4.22E+37
3.19E+50		1.77E+32	1.21E+38
7.55E+21		4.19E+03	5.66E+35
1.99E+49		1.10E+31	7.54E+36
1.29E+50		7.18E+31	4.90E+37
6.97E+47		3.86E+29	2.64E+35
			2.87E+38
1.39E+43		7.72E+24	left out
3.98E+53		2.21E+35	
		2.21E+35	
7.58E+50			
MASS OF UNIVERSE			

This was all put on a spreadsheet so I could manipulate
the numbers to fine tune the mass of the Universe. After

playing with every knob, masses, numbers of stars, number of Galaxies, sizes of stars, distribution

of stars, all within reason and adding in Black Holes and heavy dust numbers, I could not come close to the theoretical 3×10^{55} number. **That is**, until I added in a 3D Ether calculation based on the mass of electrons and protons in an 8 electron cube with a single proton in the center. Taking 1 cubic meter and dividing it into 64 smaller cubes each with a proton in the center gives the magic number. Because of electron sharing with stacked cubes. 3.53E+55. The bottom line here was to show that an Ether must exist to help equal the estimated mass of the Universe. The total worksheet was a Proportional worksheet based of the most accurate numbers available. In other words, every Star in the Universe that was the same size as our Sun was assigned 8 Planets and the Mass there of. This was continued proportionally until all Star Mass was accounted for, then came moons and Asteroids and eventually Dust. Trying to account for everything.

TIME DILATION

Where to start? Since many people claim this has been Tested several times, showing Time Dilation, let's Examine closer. If an Engineer did a test, and the Clock in motion showed a difference from a ground based clock, an Engineer would instantly think that something has affected the clock. Then all possibilities would be listed and reviewed, some possibilities could be eliminated. But for the rest, tests would be devised to simulate original As Tested conditions, even if it meant repeating the original test, but under more controlled conditions and better test equipment. Heat, Cold, Vibration, atmospheric pressure. Humidity, up/ Down, left/right movement, radiation, electromagnetic waves and magnetism. After all this, the clock would be specially prepared to filter out these possibilities (noise) and the tests ran again, paying close attention to in test operations, and any extra data available, like the Flight Recorder data for deviations to course, speed and altitude, all of which could affect time of flight. Because this was not done in any of the supposed tests proving Time Dilation, it would be Equally valid to assume something else was affecting the clocks. This is what Engineers do, no stone unturned. But it gets worse; Let's take Gravitational Dilation. This is easy, meaning the higher you go, the faster clocks run. In the atmosphere, the less atmospheric pressure is Easy to Test, put the clock in a pressure chamber and increase and decrease the pressure and see how the clock is doing. Was this done? No! I am not in favor of playing with Clocks. Below the Ocean, at sea level, on top of a Mountain, 12 AM is 12 AM. And if it is not, FIX YOUR CLOCK. Time Dilation, FIX YOUR CLOCK. Imagine the complexity if you needed a computer to tell you the Time by how fast you where moving and by what your altitude is. But here is the real reason why Time Dilation DOES NOT EXIST. As you Start moving your Clock starts slowing down at an ever increasing rate as speed increases. The problem is: Speed = Distance/Time. But, the faster you go the more that the Clock slows down(Time Dilation), the more the Clock slows down the higher your Speed increases because covering the same distance in less time = more Speed. No need to accelerate, It's an endless loop (Or a no solution Equation) More Time Dilation = More Speed=More Time Dilation=More Speed, etc.,etc.,etc. And just to make this perfectly clear; if a Spaceship flies a mile in a minute, it appears to be going 60 mph, But Time Dilation said the clock slowed down. Instead of a minute to go a mile, it only took 55 seconds, so the speed was really 65 mph, but if the speed was higher, Dilation would cause the clock to have slowed even more, which would then cause the speed to go up, an endless loop. By this logic, I guess Infinity Speed is possible. And guess who didn't see this little loop problem? Or should it be called the Time Dilation Paradox? Wouldn't One Universal Clock be so much easier. Here that?, National Academy of Physics. If you allow people to create their own Universe, it makes it increasing hard to find the real Truth. I am not a big supporter of the Nobel Prize Committee, but they never awarded the Nobel Prize to either General Relativity, Or Special Relativity.

www.ingramcontent.com/pod-product-compliance
Lightning Source LLC
Chambersburg PA
CBHW030911180526
45163CB00004B/1783